AF679518

Space Exploration

Space Exploration

Sloane Montgomery

CONTENTS

First Printing, 2025

Introduction

Humanity has always been driven by the urge to explore the unknown. From walking on the moon to delving deep into our solar system, and maintaining a continuous presence aboard the International Space Station, our journey into space has been extraordinary. But the horizon beckons us to take our next giant leap—the challenge of sending humans beyond low Earth orbit, to asteroids, and eventually to Mars.

Imagine the revolutionary strides in planetary science if astronauts could study firsthand a celestial remnant from the birth of our solar system. Such an object might hold secrets about the earliest epochs of planetary formation. Achieving this level of exploration is not just about locating a needle in a haystack—it's about pinpointing a single thread on that needle. The precision and adaptability required make humans indispensable for this endeavor.

A mission to an asteroid represents a crucial stepping stone. Although it would span only a fraction of the distance to Mars, it would allow us to develop critical skills and technologies, from navigation to habitat sustainability, that are vital for deeper space exploration. This step is not merely a practice run; it's a transformative stage in our journey toward the red planet.

NASA's ambitious goal of sending astronauts to an asteroid by 2025 marks a pivotal milestone in our quest. Asteroids, as primordial relics of our solar system, offer unparalleled opportunities to uncover its history and the processes that shaped planetary bodies. By reaching them, we unlock doors to scientific insights that could redefine our understanding of the cosmos—and bring humanity ever closer to stepping foot on Mars.

1

The History of Space Exploration

The study of space, or space science, is a relatively new and continually evolving field. While humans rely on their senses—sight, hearing, taste, smell, and touch—to observe the world around them, science expands our perception far beyond these natural limits. Telescopes allow us to see across unimaginable distances, microscopes reveal intricate details invisible to the naked eye, and tools like scales detect subtle changes imperceptible to our senses. Each instrument serves a unique purpose—some designed to tackle specific challenges, while others unveil universal processes and laws governing the cosmos.

When it comes to space exploration, an interdisciplinary effort involving scientists, engineers, computer experts, and even medical doctors has been pivotal. Together, they seek answers to some of humanity's most profound questions. Their research delves into understanding the Sun—its structure, the phenomena of solar events, and surface renewal processes previously hidden from view. Planetary orbiters capture breathtaking images of remote worlds, offering glimpses of alien surfaces. Data collected from these missions unravel mysteries surrounding planetary atmospheres and climates. Beyond our solar system, the study of celestial phenomena, such as the

X-ray radiation emitted by exploding stars in distant galaxies, continues to expand our knowledge of the universe.

The vastness of outer space, stretching beyond Earth's protective atmosphere, has captured human imagination for millennia. Ancient civilizations gazed in wonder at the Sun, Moon, and stars, drawing meaning from their patterns and cycles. Yet it was only in recent history that humanity achieved direct contact with space, breaking free from the constraints of Earth's gravity.

The first monumental breakthrough occurred on October 4, 1957, when the Soviet Union launched the world's first man-made satellite, **Sputnik 1**. This small sphere, equipped with rudimentary radio transmitters, marked the dawn of a new era—humanity's entry into the "space age." From that moment, the possibilities seemed limitless.

Following this historic milestone, nations around the world rapidly advanced their space programs. The United States and the Soviet Union, in particular, spearheaded the early space race, competing to demonstrate technological prowess and achieve firsts in space exploration. Since then, numerous countries have joined in this endeavor, launching hundreds of Earth-orbiting satellites and sending spacecraft on missions to explore other planets within our solar system.

Today, the exploration of space is no longer confined to distant observation. It is a tangible and dynamic frontier that humanity continues to push forward. As we move further into the present and the near future, the potential to utilize space's vast resources for the benefit of all people becomes increasingly clear. From uncovering the secrets of the cosmos to developing new technologies that transform life on Earth, space exploration remains one of humanity's greatest endeavors—an adventure that is only just beginning.

2

The Benefits of Space Exploration

As humanity faces an array of future challenges—ranging from environmental crises to developmental hurdles—space technology has emerged as an indispensable tool for ensuring the sustainability of human life. According to Ismatullo Zhobirov, director of the Space Research Centre of CIS countries, decades of progress in space technology have revealed its transformative potential. It serves as the gateway to a new world, offering solutions to critical issues and shaping human culture in profound ways.

Space technology already plays a pivotal role in everyday life, revolutionizing fields such as production, transportation, and global communication. Beyond these practical applications, it serves as humanity's most valuable instrument for understanding the universe and our place within it. From addressing the global food crisis to tackling energy shortages, environmental degradation, and disparities between developed and developing nations, advancements in space exploration have begun to unify human progress on a global scale. They promote the equitable development of essential resources such as food, shelter, education, and technological innovation—laying the groundwork for a brighter future.

To illustrate the multifaceted benefits of space exploration, consider these five key steps:

Step 1: Space exploration has the potential to counteract the effects of Earth's gravity, which can have adverse impacts on human health and aging. Research conducted in microgravity conditions has already provided valuable insights into the functioning of the human body, leading to advancements in medical science and therapies on Earth.

Step 2: The spirit of exploration, honed through thousands of generations, is integral to human identity. Space exploration embodies this drive, channeling humanity's historical experiences in hunting, agriculture, and discovery into new frontiers. It reawakens our collective ingenuity and inspires future generations to dream boldly.

Step 3: Space exploration represents a long-term, "moonshot" approach to addressing societal challenges. It connects groundbreaking scientific discoveries with broader societal progress, encouraging us to tackle ambitious goals while embracing the co-benefits that arise from such endeavors.

Step 4: By opening up new opportunities for trade and resource utilization, space exploration fosters wealth creation and addresses global poverty. The equitable distribution of these advancements has the potential to reduce disparities and promote social equity worldwide.

Step 5: Perhaps most significantly, space exploration links humanity to the greater expanse of the universe. It represents the next giant leap in our species' history—one that not only satisfies our innate curiosity but also reinforces our interconnectedness as a global society.

The question of what space exploration offers humanity is both timely and thought-provoking, especially as we reflect on the 50th anniversary of the first human landing on the Moon on July 20,

1969. That landmark event symbolized the triumph of human ingenuity and cooperation. Now, as we look ahead to the next half-century, the renewed enthusiasm for open and collaborative space exploration promises to yield benefits far surpassing the limited scope of focusing solely on national security aspects in space.

Space exploration is not just about venturing into the unknown; it is about leveraging those journeys to address the pressing issues facing humanity. By embracing the possibilities of the cosmos, we position ourselves to build a more resilient, equitable, and prosperous future for all.

3

Challenges and Risks in Space Exploration

The pursuit of space exploration is fraught with challenges and risks, requiring ingenuity and perseverance to overcome. Among these challenges is the development of long-term, self-sustaining technologies capable of functioning beyond the boundaries of Earth's resources. In the harsh environment of space, where resupply missions are limited and costly, sustainability becomes a matter of survival.

One significant hurdle is posed by the Van Allen radiation belts—zones of intense radiation encircling Earth. These belts expose spacecraft and astronauts to dangerously high radiation levels, requiring innovative shielding technologies and strategies to mitigate health risks. This challenge underscores the complexity of operating in the unforgiving realm of space.

Another pressing issue is the lack of a robust motorized space traffic control system. With thousands of satellites, spacecraft, and debris orbiting Earth, the risk of collisions is ever-present. Unlike terrestrial vehicles, spacecraft lack effective braking systems, making precise timing and coordination critical. Minor errors in timing or trajectory can have catastrophic consequences, further complicating the already daunting task of managing space traffic.

Before 1957, Earth's orbit was a pristine and peaceful expanse. However, the launch of the first artificial satellite, **Sputnik 1**, marked the beginning of humanity's presence in space—a presence now accompanied by growing amounts of debris. Today, Earth's orbit is a crowded arena of operational satellites, communication channels, and space junk. This ever-increasing debris poses a significant threat, as even small particles traveling at high velocities can cause severe damage to spacecraft. Alarmingly, a single piece of space debris can remain in orbit for centuries, perpetuating the risk of collisions and complicating future missions.

As we look to the future, the challenges of space exploration extend beyond debris management. Infrastructure capable of supporting the demands of advanced operations is essential. For example, satellite internet systems aim to provide global connectivity at lower installation costs, yet they also require sophisticated networks and robust energy-efficient technologies. Recycling and minimizing energy consumption in space are not just desirable but necessary—yet achieving these goals is anything but simple.

Further complicating matters is the need to anticipate the eventual introduction of intelligent, autonomous machines into space. These machines, while potentially transformative, will require careful design to navigate the complex environment of space. Additionally, the cosmos itself is an intricate and dynamic system, teeming with signals that are reflected, deflected, transmitted, and received. Managing this cacophony of information will be an ongoing challenge for future generations.

Space exploration is not without its dangers, but it is also a testament to human resilience and innovation. As we strive to overcome these hurdles, we reaffirm our commitment to advancing science and technology for the benefit of humanity. Every challenge met and every risk mitigated brings us closer to realizing the boundless possibilities that space exploration holds.

4

Current and Future Missions

The exploration of Mars has been a cornerstone of modern space science, and the journey is far from over. As humanity pushes the boundaries of what is possible, both current and future missions continue to unravel the mysteries of the Red Planet and pave the way for even more ambitious endeavors.

In the 2020s, NASA has envisioned groundbreaking missions that promise to take Mars exploration to the next level. Among them is the **Mars Astrobiological Lander (MSL)**, an in-situ mission dedicated to exobiology and life detection, alongside a highly anticipated sample return mission. These missions aim to delve deeper into Mars' ability to support life, past or present, and bring back samples to Earth for detailed analysis—an achievement that would mark a monumental leap in planetary science.

Mars rovers have already laid a solid foundation for these efforts. On January 25, 2004, the twin rovers **Spirit** and **Opportunity** successfully landed on Mars. As the first mobile scientific laboratories to explore the planet, their mission was to piece together the history of Martian water—a vital clue in understanding the planet's habitability. Over the years, these rovers sent back invaluable data, shedding light on the processes that shaped Mars' surface and revealing a once water-rich environment, forever altering our view of the Red Planet.

Spirit, despite facing insurmountable challenges, became an emblem of perseverance and discovery. After six years of unprecedented exploration, Spirit ultimately transitioned to a stationary research platform when she became trapped in a sandtrap. Her contributions, however, remain a testament to the resilience of space missions and their ability to overcome the harsh conditions of Mars.

Meanwhile, Opportunity continued her journey, accomplishing extraordinary feats such as discovering a meteorite on the Martian surface—an event that highlighted the dynamic nature of Mars as a destination for exploration and discovery.

On November 26, 2011, NASA launched the **Mars Science Laboratory (MSL)** mission, carrying the rover **Curiosity**, the most advanced robotic explorer to date. Curiosity, aptly dubbed the "ultimate puzzle-solver," represented a new era in Mars exploration. Designed to assess whether Mars ever harbored conditions suitable for microbial life, Curiosity's mission also aimed to investigate the planet's abiotic environment and its potential past habitability.

Curiosity successfully landed on Mars on August 5, 2012, following nearly a decade of meticulous development and testing. Since then, the rover has been conducting groundbreaking research, analyzing soil, rocks, and the atmosphere to uncover the secrets of Mars' ancient environments. Its findings have deepened our understanding of the planet, setting the stage for future manned missions to Mars.

As humanity prepares for its next steps, the conceptual **Mars Sample Return mission** represents one of the most ambitious projects yet. The mission envisions a collaborative effort between a surface lander and a rover to collect and transport Martian samples back to Earth. Such a feat would not only provide unparalleled insights into the planet's geology and potential biosignatures but also serve as a critical precursor to eventual human exploration.

The story of Mars exploration is a testament to human ingenuity and curiosity. From the pioneering efforts of Spirit and Opportunity to the ongoing investigations by Curiosity, each mission has brought us closer to answering one of the most profound questions: Could Mars have ever supported life? As we look toward the future, the missions of the 2020s promise to bring us even closer to uncovering the mysteries of Mars and, perhaps, the origins of life itself.

5

The Role of Technology in Space Exploration

Advancements in space exploration are the result of a profound interplay between human ingenuity and the natural laws of the universe. At the heart of this progress lies a crucial question: How can technology facilitate more rapid access to the underlying principles of nature governed by the laws of physics? This concept of "directness" refers to the ways in which technology bridges the gap between theoretical understanding and practical application—enabling us to unlock nature's secrets more efficiently.

The sciences of astrophysics and particle physics are deeply intertwined with this idea, as they explore the fundamental forces and particles that govern the universe. Through the lens of these disciplines, technology becomes more than a tool—it becomes a catalyst for discovery, allowing humanity to push the boundaries of scientific and technological development.

One of the essential functions of technology is its ability to recognize patterns and channel them into actionable directions. This capability transforms abstract natural or physical attributes into practical, useful components. Technology, in this sense, serves as the interpreter of nature's language, translating complex phenomena into tangible solutions that benefit humanity.

When considering future advancements in space exploration, the role of technology becomes paramount. It acts as a powerful engine for progress, driving economic growth, societal advancement, and scientific breakthroughs. From spacecraft capable of traversing the vast distances of the cosmos to advanced sensors that analyze the composition of distant planets, technological innovation underpins every aspect of our journey into space.

However, the consequences of stagnation in technological development are dire. A slowdown in progress could lead to economic stagnation, reduced innovation, and missed opportunities to address global challenges. This underlines the importance of sustaining momentum in technological advancements—not just for space exploration, but for humanity as a whole.

To ensure that technology continues to propel us forward, we must consider several factors. First, we must prioritize investment in research and development, fostering a culture of innovation that encourages bold ideas and transformative solutions. Second, we must leverage emerging technologies—such as artificial intelligence, robotics, and quantum computing—to tackle complex challenges and expand our capabilities in space. Finally, we must adopt a collaborative approach, uniting scientists, engineers, and policymakers from around the world to share knowledge, resources, and expertise.

The relationship between technology and space exploration is a symbiotic one. As technology advances, it opens new frontiers for exploration, and as we venture further into space, we gain insights that inspire the next wave of innovation. This dynamic interplay ensures that our journey into the cosmos is not just a pursuit of knowledge, but a testament to the resilience and creativity of the human spirit.

As we gaze toward the future, the role of technology in space exploration will only grow more significant. By embracing its potential, we can chart a course toward discoveries that were once deemed impossible—and secure a legacy of progress for generations to come.

6

International Collaboration in Space Exploration

Space exploration offers unique benefits that engage people from all walks of life, transcending national boundaries and uniting humanity in the shared pursuit of knowledge. The prospect of establishing crewed bases on extraterrestrial sites, such as the Moon, in the coming decades has the potential to captivate the imaginations of billions. These future milestones could inspire educational opportunities akin to those sparked by today's robotic exploration of the universe. As we reach for the stars—*ad astra*—we also strengthen the bonds that connect us here on Earth.

International collaboration plays a critical role in addressing the multifaceted challenges of space exploration. Advanced technological and engineering solutions, developed through partnerships between nations, can also have transformative effects on Earth. Such cooperation brings together countries with varying levels of development, fostering inclusivity and shared progress. By uniting global resources and expertise, these alliances enable us to tackle pressing issues that extend far beyond the limits of our atmosphere.

The path forward is not without obstacles. Differences in national space policies, competing interests, and the potential for geopolitical tensions may result in disputes that require resolution.

Open dialogue and good-faith negotiations among members of the international community will be essential for overcoming these challenges. History demonstrates that all complex problems—whether in the past, present, or future—have driven diverse and widespread cooperative efforts, ultimately leading to effective solutions.

The importance of international collaboration in space exploration becomes even clearer when viewed through a historical lens. Early in the 20th century, technological advancements gave humanity the means to inflict unprecedented devastation during two World Wars. These dark chapters of history underscore the importance of pursuing progress that uplifts humanity rather than divides it. Today, the pursuit of peace and global harmony stands a far better chance of success through initiatives that prioritize international partnerships, including those aimed at exploring space.

Prominent examples of global cooperation can already be seen in initiatives addressing Earth's most pressing challenges. The Intergovernmental Panel on Climate Change (IPCC) fosters global understanding and action to combat climate change, while the World Health Organization's efforts during the coronavirus pandemic illustrate the power of collective action in the face of adversity. These successes serve as models for the kind of international collaboration that will be required to make space exploration a truly universal enterprise.

As we push the boundaries of human achievement, space exploration offers an unparalleled opportunity to strengthen the bonds between nations. By working together, we can achieve far more than any single nation could accomplish alone. The technologies, knowledge, and insights gained through these partnerships not only advance our understanding of the universe but also serve as a beacon of hope for a more united and prosperous world.

7

The Commercialization of Space

As humanity ventures deeper into space, the question of commercialization takes center stage, blending economic ambition with technological ingenuity. Among the many challenges of living away from Earth's protective magnetic field, there are already demonstrable solutions, yet they demand a global economic system far more advanced than today's mix of structures. A capitalist framework appears well-suited to frontier settlement—not only for its appeal but because it facilitates the development and institutionalization of private property through advanced technology, even amid industrial imperfections.

The governance of space activities is guided by international treaties, such as the **United Nations' 1967 Outer Space Treaty**. This treaty assigns accountability to nation-states for any satellite launched into space, stipulating that "the activities of non-governmental entities shall require authorization and continuing supervision." Essentially, nation-states retain considerable decision-making power over private enterprises operating in space. Furthermore, the treaty boldly outlines provisions for the exploitation of celestial resources—including those on the Moon—but reserves these rights exclusively for nation-states, excluding non-governmental agencies under their jurisdiction.

When discussing the privatization of space, public attention often turns to prominent entrepreneurs like Elon Musk and Jeff Bezos. Their bold visions for interplanetary travel and colonization have captured imaginations worldwide. Yet, interstellar tourism, though captivating, represents only a fraction of the broader potential for economic activities in space. These entrepreneurs are often seen as visionaries, with Musk, for instance, emphasizing the importance of colonizing Mars to "safeguard the long-term future of humanity." While some dismiss their aspirations as eccentric billionaire fantasies, these endeavors reflect a more profound narrative about humanity's future.

Critics of Martian colonization question what a planet like Mars can truly offer to a growing and diverse population with varying and indefinite ambitions. The allure of Mars, in the eyes of skeptics, lacks commitment and practicality when viewed through the lens of public discourse. However, this perspective overlooks the transformative potential of space commercialization as a whole.

Commercial interests in space extend far beyond planetary colonization. They encompass a wide range of industries, including satellite-based communications, resource extraction from asteroids, and the potential establishment of manufacturing hubs in microgravity. These opportunities promise to reshape global economies and inspire new technological frontiers. The burgeoning "space economy" offers a glimpse of a future where space commercialization isn't merely a pursuit of the wealthy but a cornerstone of global progress.

As humanity navigates this complex landscape, the balance between innovation, governance, and equitable access becomes paramount. The commercialization of space presents unparalleled opportunities, yet it also demands careful stewardship to ensure that these benefits are shared broadly. Through continued collaboration and a commitment to responsible exploration, humanity can unlock the full economic and scientific potential of the final frontier.

8

Space Tourism

The concept of "space tourism" often evokes images of casual, luxurious trips to the cosmos, but the reality is far more profound. The potential market for private astronauts, or so-called "space tourists," is composed of a select group of resolute, skilled, and highly motivated individuals. For decades, obtaining opportunities for space travel through traditional government-sponsored, goal-oriented programs proved a major barrier for private candidates. High costs, coupled with the lack of access to specialized ground-based training programs, presented roadblocks that few could overcome.

Despite these challenges, those who have ventured into space as private astronauts have displayed extraordinary dedication, passion, and determination. Far from seeking a mere thrill, these individuals approached their journeys with a deep appreciation of the personal achievement it represented and the inspiration it could provide to others. Their missions, while often overshadowed by the term "tourism," have brought significant attention to the possibilities of space travel.

Much has been written about the expanding market for space tourism. With a global population of over six billion people, many Earth-bound residents dream of experiencing an out-of-this-world adventure. If not for the high costs and potential long-term safety risks, the number of eager space travelers would undoubtedly grow astronomically. For those fortunate enough to participate, space travel offers one of the most unique and elite experiences imaginable.

Currently, space tourism is the most publicized form of space travel, but the term itself is somewhat misleading. It implies a frivolous, leisure-oriented pursuit, overlooking the serious and multifaceted nature of private astronaut activities. The bulk of individuals utilizing commercial space facilities are not tourists in the traditional sense; they include professional astronauts, researchers conducting scientific experiments, educators spearheading space-related projects, and those overseeing the deployment or management of scientific payloads.

In truth, tourism represents only a small fraction of the private astronaut market. However, the perceived glamour of space travel, fueled by media depictions and a fascination with "celebrity astronauts," has shaped public opinion around the concept of space tourism. This narrative often diminishes the rigor and intent behind these missions, reducing them to casual adventures rather than groundbreaking endeavors.

Space tourism, while still in its infancy, represents the beginning of a new era in space exploration. Visionary entrepreneurs like Elon Musk and Jeff Bezos have championed the commercialization of space, expanding its reach to include private individuals alongside traditional government programs. As this industry grows, it holds the potential to democratize access to space, inspire future generations, and transform humanity's relationship with the cosmos.

The road ahead is not without challenges, but the promise of space tourism is undeniable. Whether for scientific discovery, educational outreach, or the sheer wonder of experiencing the vastness of space, private astronauts continue to push boundaries, proving that space travel is not merely a pursuit for the elite, but an aspiration for all of humanity.

9

The Search for Extraterrestrial Life

The search for extraterrestrial life has long been one of humanity's most profound scientific pursuits. By the early 1980s, however, interest in this quest had begun to wane. The Viking landers, which had successfully operated on the surface of Mars, had fulfilled their primary mission: to search for signs of life. While their findings were groundbreaking, they also raised complex questions that tempered the initial excitement.

Dr. James Lovelock, a pioneering scientist, proposed a fascinating hypothesis about Earth's ability to support life. He suggested that our planet operates as a near-regulatory system, maintaining conditions at thermodynamic limits that are both stable and open to external energy and matter flux—primarily from the Sun. This "planetary thermostat" seemed almost too perfect, as though it had been finely tuned. The Viking landers' data added to this intrigue, revealing a significant reduction of carbon into life forms, though only a small fraction of this was biotic. The microbial reactions observed on Mars' surface, despite the high-energy environment, left scientists grappling with the possibility that such findings might never yield definitive proof of extraterrestrial life. Some even feared that exobiology might inadvertently reinforce the notion of human exceptionalism.

The journey to uncover life beyond Earth began decades earlier, in the early 1960s, when humanity launched its first planetary probes. Earth-bound telescopes had already detected intriguing phenomena on Mars' surface, recorded photographically as analog data. In an innovative approach, these photographs were distributed to the public, who were invited to highlight areas of interest. This initiative, arguably the first live, global public astronomy project, captured the imagination of many—including the renowned Carl Sagan. Sagan himself observed flickering on Mars' surface and entertained the possibility of life on the Red Planet.

With the advent of spacecraft, the search for extraterrestrial life gained momentum. Missions like the Viking landers marked the beginning of a robust and systematic exploration of Mars and other celestial bodies. These efforts have since evolved into sophisticated endeavors, leveraging cutting-edge technology to probe the cosmos for signs of life.

Today, the question of whether we are alone in the universe continues to captivate scientists and the public alike. Recent advancements, such as the use of the James Webb Space Telescope (JWST), have opened new avenues for detecting biosignatures—chemical indicators of life—on distant exoplanets. Researchers are now exploring unconventional targets, such as Hycean planets, which are larger than Earth and possess thick hydrogen atmospheres. These worlds, though vastly different from our own, may harbor microbial life in their deep oceans.

As we delve deeper into the cosmos, the search for extraterrestrial life is no longer just about finding answers—it's about understanding the conditions that make life possible and our place within the vast expanse of the universe. By the end of this exploration, we may not only uncover whether life exists elsewhere but also gain a deeper appreciation for the intricate and interconnected systems that sustain life on Earth.

10

The Future of Space Exploration

The concept of *Pax Selenarum*, or the "Peace of the Moon," envisions a transformative moment in human history—a renaissance in space exploration that begins when humanity establishes its first enduring settlement on another world. This milestone marks more than a physical relocation; it signifies an irreversible commitment to the cosmos and the expansion of humanity's sphere of influence. Much like the interconnected networks that developed on Earth, this new frontier will become a vital extension of human civilization.

The foundation of this vision rests on the principles of cooperation. History has shown that collaboration among diverse nations fosters peace and drives progress. By sharing the developments and advancements of space exploration openly with the global community, humanity can create a space-reinvigorated system that benefits all. This is not simply about "sending people to the Moon"—it is about igniting a renaissance for humanity, inspired by the establishment of a world beyond Earth's boundaries.

Pax Selenarum: A Vision for Humanity

Imagine a future where Earth serves as the vibrant core of a vast network of human communities, spanning from solar-power satellites orbiting above to outposts scattered throughout the cosmos. In this interconnected web, the only limitless trend in history—knowledge without boundaries—guides our collective efforts. On clear, moonlit nights, inhabitants of distant settlements will peer through telescopes, gazing back at Earth with admiration. Meanwhile, those on Earth and its Moon will explore the universe with clarity previously unimaginable.

This future envisions experiments conducted in microgravity labs, revealing groundbreaking insights into physics, biology, and materials science. High-energy physics experiments, performed in the unique conditions of space, will unlock new realms of understanding. Together, this network of activity will revolutionize life on Earth by harnessing the limitless resources of space—providing energy, materials, and innovation to meet the needs of a growing population.

The Future of Space Exploration

Call it reconnaissance, economic expansion, or even space tourism—humanity stands on the brink of establishing its next great beachhead, one far beyond the oceans of Earth. A new wave of pioneers, entrepreneurs, and explorers will venture into the unknown, building a foundation for future generations. These settlements will not only seek resources and scientific knowledge but will also push the boundaries of human experience, redefining what it means to live and thrive as a species.

Over time, humanity will venture further into the cosmos, propelled by the same spirit that once drove us to explore uncharted lands on Earth. These bold steps will carry us closer to a future where the vastness of space becomes our playground, our laboratory, and our home. As we continue to expand our presence among the stars, we reaffirm our identity as an adaptive and pioneering species, united by curiosity and the drive to explore.

11

Ethical Considerations in Space Exploration

As humanity prepares for the immense challenge of establishing space settlements, ethical considerations must take center stage. The success of these endeavors hinges not only on solving engineering and scientific problems but also on addressing the complex moral questions that arise when launching human populations beyond Earth's confines.

One of the most pressing physical challenges for building space settlements is ensuring a benign geomagnetic environment. Space settlers will face exposure to low gravity and radiation, conditions that have unknown long-term effects on biological systems. Among the uncertainties is the potential impact of "low-g superfertility"—a phenomenon that could disrupt terrestrial norms of reproduction and development. Given these uncertainties, it is imperative to include a diverse cross-section of humanity in the space program, encompassing individuals who are older, infirm, sterile, disinterested, or part of marginalized communities. This inclusivity reflects the ethical obligation to ensure that all demographics have a stake in humanity's future beyond Earth.

Decades of experience, drawn from five significant studies on space biospheres, provide valuable insights into the practical and ethical dimensions of constructing humanity's first "nation beyond nations." These studies underscore the importance of planning for the ethical operation of such settlements, ensuring that they serve as inclusive, just, and sustainable societies. However, history warns of the potential risks: the presence of individuals with destructive tendencies—whether narcissists, bioterrorists, or doomsayers—could implicate all of humanity in any fallout, emphasizing the critical need for rigorous governance and accountability.

The establishment of orbital space colonies forms the backbone of effective space settlement implementation. These colonies provide opportunities to test and refine the technologies and systems needed for long-term habitation. Currently, the most active area of biological space settlement planning focuses on addressing key challenges such as costs, the development of lab-grown meat, animal husbandry, and efficient waste recycling. Each of these components presents significant ethical dilemmas. For example, lab-grown meat raises questions about resource allocation and energy use, while waste recycling systems must be designed to minimize environmental impacts both in space and on Earth.

Space analog studies conducted in suborbital, earthbound environments offer critical data for these models. By simulating conditions of isolation, resource scarcity, and confinement, these studies shed light on the psychological and social challenges that future settlers may face. They also highlight the importance of designing settlements that foster cooperation, resilience, and a shared sense of purpose.

Beyond the practicalities, the ethical considerations of space exploration extend to the larger question of humanity's role in the cosmos. As we expand into space, we must confront issues such as the preservation of extraterrestrial environments, the equitable distribution of resources, and the potential cultural and societal shifts that space settlement may induce. Failure to address these concerns could not only jeopardize the success of these missions but also risk a moral regression—a "new coloration to a second dark age."

Ultimately, space exploration offers humanity an unparalleled opportunity to redefine its ethical boundaries and aspirations. By building settlements that reflect the best of our values—equity, sustainability, and inclusivity—we can ensure that the expansion of humanity into the cosmos is not just a technical triumph but a moral one as well.

12

Space Exploration and Climate Change Research

Humanity's horizon for space exploration continues to expand, from the Moon to Mars and beyond. As we prepare to traverse the vast expanse between terrestrial planets and icy celestial bodies, a critical question emerges: How can we transform our search for life on Mars and the study of tenuous atmospheres into a driving force for advancing climate change science? By coupling disciplines such as Heliophysics, systematic space exploration, and Earth observation, we unlock powerful synergies that hold immense potential to advance both fields. This integrated approach not only strengthens our National Endeavor but also showcases science as a global investment while positioning humanity as a leader in technological innovation.

The natural coupling of space exploration and climate change research highlights their shared challenges and opportunities. Both fields grapple with deciphering the significance of natural events, understanding climate uncertainties, and unraveling mechanisms of action that span multiple spatial and temporal scales. These challenges require new methodologies and interdisciplinary efforts to push the boundaries of knowledge.

Space exploration, with its focus on the systematic study of celestial bodies and asteroids, yields data that often defies expectations, encouraging fresh qualitative interpretations. Similarly, climate change science benefits from a more refined resolution of observations in space and time. Yet, limitations persist: either climate events are not being observed with sufficient precision or their magnitude places them beyond easy comprehension. This conundrum is mirrored in space exploration, where phenomena unfold across diurnal, annual, and even longer timescales, complicating our ability to draw definitive conclusions.

To address these challenges, we propose dedicated research investments that amplify the synergy between these fields. For instance, Heliophysics—the study of the Sun's interactions with the solar system—offers valuable insights into energy flux and atmospheric dynamics that can be applied to understanding Earth's climate systems. Similarly, techniques developed for space missions, such as high-resolution imaging and spectroscopy, can be adapted to monitor climate change with greater accuracy.

Moreover, this partnership encourages a deeper appreciation for the interplay between planetary science and Earth science. The illuminating of Mars' potential for life or the exploration of icy bodies with tenuous atmospheres offers analogs to processes occurring on Earth. These comparisons enhance our ability to predict large-scale climate phenomena and develop strategies for mitigation and adaptation.

The synergy between space exploration and climate change research not only advances scientific understanding but also inspires global collaboration. By treating these two domains as interconnected, we foster a holistic approach to solving some of humanity's greatest challenges. As we venture further into space and deepen our understanding of Earth's climate, we reaffirm the importance of innovation, cooperation, and the limitless potential of science to shape a better future.

13

The Role of Space Agencies

Space exploration relies heavily on the vision, funding, and coordination of national space agencies. In the last year, the combined annual budgets of these agencies across the globe totaled approximately 75 billion USD or less. Among these, NASA's budget stands at slightly over 20 billion USD—significantly smaller than the historic peak funding levels seen during the Apollo era.

The Apollo program, one of NASA's crowning achievements, spent 25.4 billion USD by 1967 to send a dozen astronauts to the Moon. When adjusted for inflation, this figure exceeds 150 billion USD. By comparison, more recent programs, such as NASA's proposed Ares V and Constellation program in 2008, were estimated to require anywhere between 230 billion and 380 billion USD. This stark contrast highlights the challenges faced by modern space agencies. Securing the necessary political consensus and public support for large-scale exploration initiatives has grown increasingly difficult, particularly as societies grapple with social inequities, poverty, and the urgent need for global action on climate change.

Despite these challenges, government space agencies play an indispensable role in advancing humanity's presence in space. Without their foundational efforts, space programs would be far more limited than they are today. Agencies like NASA, the European Space

Agency (ESA), Roscosmos, and others drive innovation, conduct scientific missions, and establish critical infrastructure that benefits both space exploration and life on Earth.

In recent years, the collaborative nature of space exploration has grown significantly. International partnerships have been instrumental in major projects, such as the construction and operation of the International Space Station (ISS), where researchers from diverse nations work together in harmony. These collaborations underscore the shared goals of scientific discovery and technological advancement, transcending geopolitical boundaries.

The private sector has also emerged as a vital contributor to the space economy. Scientific and commercial organizations are working on space-related projects in a variety of capacities. The increasing miniaturization of satellite components, combined with the decreasing cost of launches, has enabled private enterprises to develop satellite solutions for communication, Earth observation, and more. This democratization of space access has paved the way for innovative technologies that improve everyday life, such as global internet coverage and environmental monitoring systems.

However, accelerating humanity's journey deeper into space will require substantial investments. The costs of sending robots to other planets, maintaining the ISS, and developing advanced spacecraft remain significant. Achieving the next milestones, such as manned missions to Mars or permanent lunar settlements, will demand unprecedented levels of funding, innovation, and collaboration.

As humanity strives to extend its reach into the cosmos, space agencies continue to serve as the backbone of exploration efforts. They provide the expertise, infrastructure, and coordination needed to tackle the immense challenges of space. By fostering partnerships with private companies and international collaborators, they amplify their impact, ensuring that the benefits of space exploration extend to all of humanity.

14

Space Exploration and National Security

The United States' space policy outlines an ambitious goal: returning to the Moon as a stepping stone to Mars and beyond. The Moon provides a fiscally manageable platform for testing technologies, refining strategies, and preparing for the next great leap in human exploration. By focusing first on the Moon-Mars pathway, we can establish a foundation for habitability and exploration without immediately delving into the complex national security objectives of space. Over time, advancements in spacecraft design and medical risk management may eliminate many of the transportation and health-related obstacles that currently hinder Martian exploration.

Long-term planning for space exploration also involves envisioning ways to sustain operations on Mars, including a hypothetical continuous Mars 24/7 schedule. It may even incorporate military contingents equipped to adapt to hostile extraterrestrial environments. However, past experiences with similar endeavors often began with limited returns and eventually faded from prominence, underscoring the need for sustained focus and investment. Innovative technologies that transcend existing devices, along with adaptable strategies, are critical to advancing these efforts. Researchers emphasize the need for efficient returns on investment, ensuring that initiatives align with broader objectives like containing costs for the Department of Defense (DOD) and enhancing capabilities for Earth-orbit operations.

Health considerations also play a pivotal role in space exploration. Challenges such as brain injury, impaired growth, and psychological impacts like low-depth-of-field depression must be addressed for long-term habitation in extraterrestrial environments. Prelinguistic communication difficulties, potentially linked to limited brain development in microgravity, underscore the need for thorough preparation and mitigation strategies to ensure human survival and resilience in space.

Space exploration represents a critical component of U.S. national power, leveraging the nation's industrial and technological base. The Department of Defense (DOD) actively supports these efforts, integrating space exploration into its broader objectives and fostering collaborations across public and private sectors, as well as with international partners. Industry involvement, through selected commercial entities, remains essential, particularly as Earth orbit, the Moon, and Mars represent key strategic destinations where private funding alone is not sufficient.

The Air Force plays a leading role in extending the United States' global reach through space initiatives. Its activities complement broader DOD research programs, drawing from investments in both space science and defense. These programs are bolstered by contributions from federally-funded institutions such as NASA, the Office of the Director of National Intelligence, the Defense Advanced Research Projects Agency (DARPA), the National Reconnaissance Office (NRO), and others. Together, these agencies ensure the preservation and enhancement of national power through space exploration.

As the United States navigates this complex landscape, reducing administrative overhead and increasing funding for national security space programs will be vital. By streamlining operations and prioritizing investments, the U.S. can maintain its leadership in space exploration while addressing critical security needs.

Ultimately, space exploration serves as both a symbol and a cornerstone of national power. It drives technological progress, strengthens defense capabilities, and inspires global leadership. As the U.S. pushes further into the cosmos, its investments in space exploration will continue to bolster its national security and ensure a prominent role on the world stage.

Conclusion

As we look to the future of space exploration, one nation poised to play an increasingly prominent role is China. From the outset of the human space exploration era, China has prioritized manned missions, demonstrating its commitment to advancing scientific and technological progress on a global stage. This dedication positions China as a key participant in shaping the next phase of international collaboration in space. With the proposed establishment of new Science Orbit Station facilities, opportunities for cooperation abound. China, alongside other nations, could contribute to the definition, development, funding, and management of these facilities and their associated experiments, cementing its role as a vital partner in the global space community.

At the heart of current efforts, the **International Space Station (ISS)** serves as a beacon of international collaboration, scientific research, and commercial development in low Earth orbit. However, as the ISS matures, the need for a new scientific objective becomes apparent. Any serious intention to return to the Moon—and ultimately venture beyond—will require significant investment in new resources, innovative materials, and robust commercial partnerships. This step forward demands the collective efforts of nations, industries, and institutions, as no single entity can achieve these monumental goals alone.

Resource limitations in global space research necessitate a strategic approach. Priority must be given to facilities and initiatives that directly contribute to humanity's progress. Areas such as resource exploration, environmental research, and sustainable development hold the potential to yield tangible benefits for life on Earth. By focusing on projects with practical applications, space exploration becomes more accessible and relevant to nations that have been hesitant to fully support manned missions in the past. These priorities not only justify the investments required but also align space exploration with broader goals of combating poverty, addressing environmental degradation, and promoting equitable development.

The future of space exploration hinges on the principles of cooperation, innovation, and inclusivity. As nations come together to share knowledge, resources, and aspirations, we collectively advance toward a future where space exploration drives not only technological progress but also societal growth. By fostering partnerships across public, private, and international sectors, we ensure that the benefits of our endeavors extend beyond the stars—enhancing life on Earth and uniting humanity in a shared quest for discovery.

Space exploration has always been more than a scientific pursuit; it is a testament to humanity's boundless curiosity and resilience. As we stand on the brink of new horizons, the lessons of collaboration and perseverance learned through decades of exploration will guide us toward a brighter and more interconnected future, both in space and here on Earth.

Enjoyed This Book? Let Others Know!

If this book has blessed, encouraged, or challenged you in any way, I'd love to hear about it! Your review not only helps others discover this message but also encourages me to keep writing.

Would you take a moment to share your thoughts? A few sentences on what stood out to you can make a big difference.

You can leave a review on Amazon, Goodreads, or wherever you purchased this book. Thank you for being part of this journey!

www.ingramcontent.com/pod-product-compliance
Lightning Source LLC
LaVergne TN
LVHW090335210926
848480LV00004B/1425